THE SCIENCE OF
FAITH

Rev. Judah Montenegro, BSE

Copyright © 2018 Judah Montenegro

All rights reserved.

ISBN:1721636372
ISBN-13: 978-1721636372

To my lovely mother
Mercedes Armida Montenegro
This book is a testament
of your love and good parenting.
I love you and miss you very much
¡Madre bendita seas!

CONTENTS

	Acknowledgments	i
Intro	Science + Bible	1
Part I	**Time to Understand**	**Pg. 13**
1	Time; according to Physics	Pg. 15
2	Time Travel	Pg. 29
3	God's Timing	Pg. 45
Part II	**Space Matters**	**Pg. 61**
4	Space & Matter	Pg. 63
5	Lord of the Dimensions	Pg. 75
6	Inspired	Pg. 93

ACKNOWLEDGMENTS

I want thank all those who helped
me make this book a reality.

First of all I want to thank My Lord Jesus Christ

L. Maasiai Montenegro – Editor, Sister
Beq Montenegro – Illustrator, Brother
Zulema Montenegro – Graphics, Sister
Esly Montenegro - Visual Editor, Sister
Rebekah Bermejo – Support, Girlfriend
Steven Ramos – Photography, Friend

Special thanks to all my siblings who inspire me!
Steve & Melissa Montenegro
Zabdy & James Rios

And my father
Dr. Jose Roberto Montenegro

Book Release Date: June 30, 2018
My Golden Year birthdate & dads birthday.
Happy Birthday Dad!

Intro

"I believe that there is a complete truth that is strong and stands tall regardless of the examiner. This truth does not tolerate variation nor alteration to its purity."

Judah Montenegro

Science + Bible

Before we begin, I believe it is important for me to express my desire to bring forth how science is in fact a tool that can and is used hand in hand with the Bible and it's truth. I am not here to discount what science brings to light, rather to show how it has helped me to solidify my faith and what I know and understand from reading the scriptures.

Science is a word that has been around for a long time. Even it's definition has changed and evolved through the course of its trajectory. It's a word that has been tossed to and fro since the beginning of it's practice. The word *science* is mentioned two times in the King James Version of the Holy Bible. Once in the Old Testament in the book of Daniel chapter 1 verse 4, "children .... understanding *science*" the Hebrew phrase meaning "those who understand *science*." The meaning of the term here is *knowledge* and *wisdom*. The other occurrence of the word *science* is found in the New Testament in the book of Timothy chapter 6 verse 20.[1] Also meaning *knowledge*. The word *science*

---

[1] https://www.biblestudytools.com/encyclopedias/isbe/science.html

was a word that was defined as knowledge or deep wisdom of any particular topic. As time went on science became more known as the discovery and orderly classification and exposition of the phenomena and of the laws of Nature.[1] If you were to google the definition today you would get something like "the intellectual and practical activity encompassing the systematic study of the structure and behavior of the physical and natural world through observation and experiment." I personally like to define science as the study and explanation of the natural observable world around us. As a college student going through many types of physics and mathematics classes I had an open mind and allowed science to build my faith. Now I use science to strengthen my faith and the faith of others around me. I use science as a tool to help me elaborate on God's beautiful creation and to dress my understanding of God's visible world with logical terminology. I have adopted the scientific disciplines of thinking to dive into theological subjects. All for the enhancement of my faith and desire to know Him more.

---

[1] https://www.biblestudytools.com/encyclopedias/isbe/science.html

We have seen and learned that the Bible has great scientific substance. Sometimes even more that we could ever understand. Somethings go right over our head and our understanding. So, the fact that we do not understand certain things does not mean they do not exist or that it is not out there. The Bible is full of truth. It is filled with truth that is real, profound and absolute. I believe in absolute truth. In the same way, I believe truth is independent of a believer or accepter of that truth. I believe truth *stands* regardless of opinions and judgments or even one's ability to reason. Let me explain, in this mind frame it is important to acknowledge that as humans we are limited in our understanding. Science is a beautiful thing. It helps us to truly grasp and understand the world in which we live. Science is constantly growing and advancing. Scientists are constantly learning, adding and taking away to the fundamental building blocks that have been established through time. The things we now know and understand through science were not always known or accepted. Over the course of the years we have come to comprehend and build on truths that have been solidified through observation and testing.

Imagine if truth depended on a believer in order to really be true? Imagine if truth needed an acceptor in order to have credibility? Could we then call it truth? Can we then let it be a source of our fundamental values and beliefs? I believe in an absolute truth that stands confident without a need to defend itself. A truth that needs no assistance.

What is truth? Is truth relative? Can truth have different interpretations? Can truth be relative dependent upon one's perspective, angle or even point of view? I believe that there is a complete truth that is strong and stands tall regardless of the examiner. This truth does not tolerate variation nor alteration to its purity. The essence of this truth will not differ in shape, shade, or quality as a result of the lens from which it is viewed. This then, is a multidimensional truth that goes far beyond the depth of our human ability to reason and understand.

Many argue to have truth. Many argue to have found a truth. Many claim to understand truth. But in all my findings in science and religions, I have found only ONE who claims to BE truth. There are many prophets in different belief systems and religions that claim to point you to a truth. Or point you to the true way of living or true way of thinking and believing. But only ONE has made the bold claim to BE truth. Jesus Christ in John 14:6 makes an incredible claim and statement. Jesus said..., "I am the way, the truth, and the life. No one comes to the Father except through Me."[1] What a statement and a claim? To BE the WAY the TRUTH, and the LIFE. He didn't say "I bring you life" or "I bring you a way of life" but Jesus said I AM the WAY. There is no other way to live, think, or believe if you want to live, think, and believe in TRUTH and to have TRUE LIFE! Jesus says in John 10:10 that others, such as the *thief,* also known as the devil, satan and/or lucifer, comes to steal, kill, and to destroy, but, Jesus Christ has come to give LIFE. Not *just* life but LIFE in abundance![2]

I believe that this absolute truth is found in the Holy Bible. A book that has been a bestseller since it's first days printed in October 22, 1455. This book, The Holy Bible is forever known as the first book ever to be printed according to the study of Johann Gutenberg. Johann Gutenberg was the inventor of the first movable-type printing press in 1455 who

---

[1] John 14:6
[2] John 10:10

produced the first book ever to be printed: a Latin language bible, printed in Mainz, Germany.[1] The bible, which has about 35 to 40 different authors in different periods of time and various walks of life[2] such as Shepherds, farmers, tent-makers, physicians, fishermen, priests, philosophers and kings still remains one of the best kept records of history till today. Despite these differences in occupation and the span of years it took to write it, the Bible is an extremely cohesive and unified book. The Bible took over 1,500 years of writing and experiences inspired by the Holy Spirit from around 1450 B.C. (the time of Moses) to about 100 A.D. (following the death and resurrection of Jesus Christ).[3]

The Bible has countless data and scientific information we are still discovering today. The Bible talks of the origins of our universe and the creation of the earth and humanity. The Bible also talks of a worldwide flood that changed the earth in so many different ways, ways in which we are still discovering the effects on this earth. The Bible takes us on a

---

[1] http://www.gutenberg-bible.com/
[2] https://overviewbible.com/authors-who-wrote-bible/
[3] https://bibleresources.org/bibleresources/bible-facts/

journey through human history and the origins of civilization. The Bible illustrates the depths of a spiritual world where human eyes reach their limits and human understandings must lean on faith because of the depths of a higher dimensional world. A world such as the spiritual world which surpasses our limited knowledge and understanding.

In this book we will be learning about how the bible refers to time, space, and matter compared to how science refers to these subjects. We will cover time from a physical standpoint and a biblical standpoint. Is time travel possible? Is there incidences in the bible where we get a glimpse of time travel? For instance, take Moses and Elijah on the mountain of transfiguration, appearing in a time that was far beyond their own.[1] Or what about Jesus, who the Bible tells us as Lord is the same yesterday, today, and forever,[2] all words used to describe references of time. Can we say that is time travel? Or is there more to it than just traveling in time?

We will be addressing higher dimensions both, mathematically and spiritually. As a Christian and

---

[1] Matthew 17:3 , Luke 9:28-30 , Mark 9:2-9
[2] Hebrews 13:8

believer of God I believe there is a spiritual world and Science has proof of higher dimensions in this world that are sometimes hard to understand and conceptualize. This is where science depends on mathematics to truly define and make concrete the concepts of these higher dimensions and to help us understand the deeper levels of our present world. We can also say that the Bible refers to depths in this world that are hard to comprehend with our finite human intuition and natural way of thinking. For example, when the apostle Paul was awestruck he said, "I know a man in Christ who fourteen years ago—whether in the body I do not know, or whether out of the body I do not know, God knows—such a one was caught up to the third heaven."[1] Could the apostle Paul be referring to a spiritual world that God allowed him to witness? Is this a higher dimension that our brains are not exposed to on a daily basis?

The purpose of this book is to challenge the believer to think and to encourage the thinker to believe. To stretch our limited understanding of the depths of who God is. To take God out of any small bubble or box we may sometimes place Him into and

---
[1] 2 Corinthians 12:2

to realize that there is more to this Christian faith than we allow ourselves to become comfortable with. Hopefully, this book provokes your thoughts and your spirit enough that it takes you to a place where you are hungry to learn more about the complexity of God and the depths of His kingdom. My truest desire in writing this book is to expand your belief and equip you with strong convictions and a deeper faith. Thus, empowering you with a faith that is steadfast, immovable, and abounding in the work of the Lord always. And this will help you to remember and never forget your work in the Lord is not in vain.[1] As you read this book my hope is for you to tap into and perceive how limitless God really is and the finite extent of our understanding.

I invite you to dive into these subjects with me and see how mathematics compliments the bible and vice versa. In doing so let's open up in a prayer so that the Lord can open up our understanding and give us the gift of knowledge when reading these findings.

---

[1] 1 Corinthians 15:58

Lord, I pray that you may help me as the author to write these truths and findings that you have given me in such a way that the readers can intake and be blessed. Blessed in a way that their faith in you is made stronger. I pray, Lord, that you may be the giver of wisdom and from your mouth we may be given knowledge and understanding.[1] I pray that our minds may be stimulated and grow closer to you in reading this book. I pray that you may deposit greater truths in the readers hearts and make them authors of great books and teachers of Your Word through your imparted and given revelations. Lord, I pray that you hear these prayers and that your kingdom come into our hearts and thoughts and Your will be done here on earth, here in our minds, here in our world as it is in heaven![2] In Jesus Name, Amen!

---

[1] Proverbs 2:6
[2] Matthew 6:10

PART I

"the past a canyon that can be climbed into and the future a mountain that can be climbed up"

- Inspired by Interstellar, *the movie*

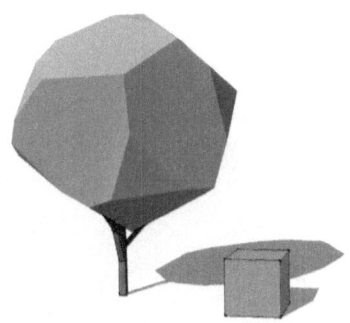

Judah Montenegro

chapter 1

"We are all a prisoner of the grips of time. Forever imprisoned in the present. Transitioning from the ever past into an eternal future."

- American Astrophysicist, Neil deGrasse Tyson

Judah Montenegro

Time to Understand

What is time? Some people call it the 4th Dimension, others call it part of the Space-Time continuum. World famous American Astrophysicist, Neil deGrasse Tyson says; "We are all a prisoner of the grips of time. Forever imprisoned in the present. Transitioning from the ever past into an eternal future." Will we ever reach the future if we are stuck in the present? Will we ever understand the past or come to realize that the past will always be a memory that doesn't exist anymore? Will we ever enjoy the now as much as we could? These are questions I have asked myself in trying to understand the concept of time. I have found that time is a concept embedded deep in the human psyche. As humans the concept of time is the norm. Something we don't even take time to think about.

But what is time really? Can science ever reach a point where we can travel into a distant future or a distant past? We can say that time is a property. A property that can be measured and calculated (we will revisit this concept with greater depth in chapter 2). Okay now that we have our brains going let me ask you this: What does time look like from a spiritual perspective? Are angels and spiritual forces subject to time? Are spirits aging? Do they ever get too old to do their spiritual duties? Do angels or spiritual creatures have an age? Have you ever thought about these things? I hope these questions provoke and awaken your curiosity and sense of wonder. I pray, of course, that these questions may only bring us to think about God and His greatness. How much more there is out there that we really don't know about, hidden in the store houses of the knowledge of God.

I want to challenge your thinking. I want you to feel my sense of wonder when I come across these conversations. Let's feel puzzled together when we talk about these topics that seem so normal in our everyday life. However, these topics have such a depth when we take some time to ponder and try to fathom the complexity therein. These questions and conversations have made me believe even more in our God and Lord, Jesus Christ.

Why do I say that? You may be asking yourself. I say that because I believe that He is the source of all the right answers. Because He is the creator of the whole physical universe that we can see with our naked eye and with the help of eye-aids such as telescopes and microscopes. In addition, the Bible tells us that God is the creator of the things that are invisible, like the spiritual world and higher dimensions. But we need faith in order to understand the things that we cannot see. Check it out; "Now faith is the substance of things hoped for, the evidence of things not seen. For by it the elders obtained a good testimony. By faith we understand that the worlds were framed by the word of God, so that the things which are seen were not made of things which are visible."[1] In other words, the things that are not seen, all the invisible things, are more real than the things that are seen. They have greater sustenance when compared to the things that are seen. Oh man, this is cool stuff. We can see that there is much depth to our faith. By faith we can understand great depths. By faith we can understand great things. By understanding great things through faith, we can reach further and deeper than we ever would be able to without faith. There are certain things we can only understand by faith. Faith in the living God, Our Lord Jesus Christ! Once we accept the fact that there is a God and He is Lord of all than the

---

[1] Hebrews 11: 1-3

doors are open to fountains of blessed understandings. Take a moment to accept this absolute truth and with your words invite Jesus into your heart if you have never done so. For God is the giver of all knowledge, wisdom, and understanding. If we have Him, we have it all. Thank you Jesus!

With that being said, Let's see keep chipping away at this topic of time. Let's see how physics defines time and what physics tells us about it.

Time; according to Physics

Before the rail transportation system arrived and was implemented in the United States of America at the turn of the 18th century, towns across the nation would keep their time defined by when local noon would occur. That is, by when the sun was directly overhead. Each town would use general, more loose, terms for scheduling, appointments, and time designation such as; dawn, midmorning, noon, early evening, dusk, and so on. This was before the invention of the portable clock, the watch. Because before then, there was no need for a portable clock, there was no need for people to know exact times. But once rail transportation was introduced people began to agree on a time, and thus, arrive to certain places at certain times. This was the birth of a synchronized era.

Check this out, according to the Macmillan Encyclopedia of Physics, the earth rotates on its axis, each longitudinal degree reaches noon at different moments in time. For every 50 miles in the Midwest United States in distance we get 1 longitudinal degree. This means it takes 4 minutes for each degree (50 miles) to have their very own noon. Thus, it takes 1 hour for every 15 degrees to have its own noon where the sun is directly above. As you can imagine, this can get very complicated and cause a lot of confusion with train departures and arrivals. So the solution to this was to establish different zones across the nation of America. It was such a great idea that it caught on and was implemented worldwide.[1] The birth of time zones.

---

[1] Rigden, J. S. *MacMillan Encyclopedia of PHYSICS* (Simon & Schuster Macmillan, New York, 1996) Vol. 4 Pg. 1610

Okay so now we see how the interpretation of time began to solidify and become an integral part of our society in order to communicate with others and be on the same page about the hour of any given moment. We can understand that this is still not yet a definition of time. But merely an explanation of this observed phenomenon of a repetitive change. What I want to further examine is a more conceptual definition of time. What is time? According to our handy dandy MacMillan Encyclopedia of Physics,

> Time is a concept "connected to the idea of ongoing and repetitive change in the physical world. Change and repetitive change is obvious. Our bodies age, but at the same time new generations are born, and their bodies age. The sun comes up each morning; seasons change and repeat; our heartbeat occurs in a fairly steady pattern, and we have an internal psychological clock that lets us judge when our heartbeat is rapid or slow."[1]

---

[1] Rigden, J. S. *MacMillan Encyclopedia of PHYSICS* (Simon & Schuster Macmillan, New York, 1996) Vol. 4 Pg. 1609

Okay so let's nerd out a bit and really take a look at this thing called time according to what physics tells us. Let's see what the folks at Oxford University Press say about time in their sixth edition of A Dictionary of Physics.

> "A dimension that enables two otherwise identical events that occur at the same point in space to be distinguished. The interval between two such events forms the basis of time measurement. For general purposes, the earth's rotation on its axis provides the units of the clock [in a day] and the earth's orbit around the sun provides the units of the calendar [in a year]. For scientific purposes, intervals of time are now defined in terms of the frequency of a specified electromagnetic radiation [in a second]..."[1]

How can time be a dimension? I will be discussing dimensions in part II so we can understand it a more fully. For now, I really like this definition of time

---

[1] Oxford University Press *A Dictionary of Physics* (Oxford University Press Inc., New York, 2009)

because it breaks it down in a simple way. I did not include, however, how it mentions Einstein's theory of Relativity. Where he abandons the concept of absolute time. Absolute time meaning "time that flows equably and independently of the state of motion of the observer."[1] In other words, Albert Einstein abandoned the concept of a certain absolute clock or absolute simultaneity that serves as reference to all and every other state of motion and being. According to his theory of Relativity, time is relative to local parameters. Which introduces the concept of Time Dilation, a phenomena of Albert Einstein's special relativity studies.

> Time Dilation: "The principle, predicted by Einstein's special theory of relativity, that intervals of time are not absolute but are relative to the motion of the observer."[2]

In other words, this is saying that absolute time is relative to your very own circumstance. As in, there is differences in a personal clock depending on the state

---

[1] Oxford University Press *A Dictionary of Physics* (Oxford University Press Inc., New York, 2009)
[2] Oxford University Press *A Dictionary of Physics* (Oxford University Press Inc., New York, 2009)

of motion and local parameters. That means that there is a slight shift in time depending on your local circumstances and motion. What is this telling us? That again, there is no such thing as an absolute time according to physics. Time is different for every individual, observer, and every being. Let me give you an example of time dilation.

According to an article I came across on space.com[1] it talks about a set of astronaut twins and how they aged differently as one spent time in outer space on his one-year mission aboard the International Space Station (ISS). The article mentions that the former NASA astronaut Scott Kelly made his older twin brother Mark Kelly even older. Scott Kelly, the younger twin, spent a total of 520 days in orbit on a few missions. During his time in space he free fell around the earth's orbit at speeds of about 17,500 miles per hour. Oh Snap! Imagine getting a speeding ticket for that one?

Here we can see that Albert Einstein's theory of special relativity actually holds through the measurements of this Time Dilation when it refers to

---

[1] https://www.space.com/33411-astronaut-scott-kelly-relativity-twin-brother-ages.html

time moving slower for objects in motion at much higher velocities in comparison to a stationary observer. Of course this theory of time dilation is a lot more dramatic and noticeable at much greater speeds such as the speed of light (186,282 miles in one second) or anything close to those speeds. However, the effects are still calculable at astronaut's velocities in low-earth orbit doing missions on the ISS or Hubble Space Telescope.[1]

Mark Kelly, the older twin, mentioned on a TV panel that he was older than his brother now, about 6 milliseconds older then he was before his brother went on his mission to space. These are approximated calculations that were used loosely by Mark but it goes to show the order magnitude of time dilation in a real life application. Talk about respecting your elders, Scott. I can imagine a new emerging commercial spaceflight company having this sales pitch, "Need more respect? Can't seem to have the respect of your loved ones? For a flat fee, Just send them into low earth orbit for a few hundred years to gain age on them and demand the proper respect you

---

[1] https://www.space.com/33411-astronaut-scott-kelly-relativity-twin-brother-ages.html

deserve." "We accept financing!" Now that's funny! As you can tell I like to have fun. Especially when we talk about these mind bending subjects.

Okay back to business, let me ask you a question. Did Scott travel 6 milliseconds to the past to make Mark even older than him? Or did Mark travel 6 milliseconds into the future to create a larger age gap and demand more respect? This is the theory of relativity folks. Is time travel possible? Does it exist? Come on, pack your bags, grab your passport and let's see what science tells us about time travel.

chapter 2

"It is through prayer that I communicate to such a powerful God. So, can it be that prayer is subject and bound by these limiting ideas and laws of time?"

Judah Montenegro

Time to Travel

Lord, help us as we take a closer look at some of these concepts and subjects. I hope you are as excited as I am about having these conversations. I've learned that there's a lot more to learn. Imagine how much we don't know? Imagine how much we don't know we don't know? In other words, imagine what we haven't even realized that we have no clue about? There's a lot out there that we may never be able to comprehend thoroughly. But we believe that our God is the true source of all knowledge.[1]

Okay, I know I have been throwing a lot of definitions and verbiage but be patient with me as we dissect these concepts and allow them to lead us into some biblical ideas.

Physics tells us that time is a property. As promised, again we see here the idea of time as a property. Just like there are other physical properties such as height, length, and width, we also have time. In 2014 a popular film *Interstellar*, directed by Christopher Nolan was released and caused a lot of

---

[1] Proverbs 2:6

chatter about these concepts and ideas regarding space-time and higher dimensions. It was indeed very impactful, especially because there was real science behind the main ideas used in the film. Nolan used American theoretical Physicist and author of "The Science of Interstellar" as a consultant to be accurate in all depictions of the film. This brings ensuring credibility to the film. I don't usually like quoting movies or getting my "knowledge" from secular films but I've done some background research in order to use it as part of our conversation. Aside from the fact that this has been one of my favorite motion pictures since 2014 because of its scientific accuracy. And of course, like any hollywood film, there was some science-fiction added to bring closure and to tie loose ends to the theoretical part of the film and the actual limitations of reality. Let's just say It's a fine work of art.

I bring this up because a character in this film, Brand (played by Anne Hathaway), describes time to be relative. She says that time can be stretched and squeezed but it cannot go backwards. Then she mentions fifth dimensional beings, or higher dimensional beings, may have the possibility to see time as a physical property. For example, the past a canyon that can be climbed into and the future a mountain that can be climbed up. Regardless if that's accurate or not, this brings some thought-provoking elements into the picture about time travel.

For a higher dimensional being, I can imagine *time* can be seen as a timeline. Set up and established. Past, present, and future all visible. I can imagine that for higher dimensional beings time is not chronological. It is more like a sea, where as a moment in time at a specific location is a drop in the sea. I can also relate it to a YouTube video where you have access to the beginning of the video as well as the end but there are a lot of moments in between that may be difficult to find especially if you are looking for *one* specific moment in time of a specific location or occurrence in this exact moment.

Later we will be breaking down lower and higher dimensions and dimensional beings but for now let's see what physics tells us about time travel. Time travel has been a topic for science fiction for decades now. It boggles the mind with mind bending scenarios. There have been movies and TV series made about traveling to the past or years into to future by entering a time machine that has all the capabilities to take you anywhere you want with a touch of a button or getting up to 88 mph with the help of the flux capacitor, for all my Michael J. Fox fans who are familiar with the Back to the Future Trilogy. It has always brought some sort of entertainment. As humans we are naturally curious individuals that like to think outside of the box. We are always inquiring and wondering what else is possible. This is the reason why this subject has become one of my favorites to think and talk about. As an Aerospace Engineer I always find it fascinating to learn about time and the mind bending concepts behind it. As a student, sitting in class learning and wondering about the depths of our God was always something to look forward to. Imagining how much more there is out there that God is yet to reveal to us 4-dimensional beings, aka humans, is a delightful journey.

Furthermore, we see that physics tells us about *The Arrow of Time*. A concept known to physicists and chemists alike. This concept states that time flows naturally in one direction. This is because there are physical and natural laws that have been observed and give testimony to this phenomenon. For example, we note the law(s) of entropy. Often interpreted as the degree of disorder or randomness in a system, in other words a measurement of order to chaos. Entropy states that a system has a definite direction of order towards disorder. Moving from order and structure to more disorder and chaos.

Allow me to give an illustration. If we drop a ceramic vessel and it accelerates towards the floor at the earth's gravitational constant traveling 32.174 ft per second, it will shatter. And the only way we can put this vessel back together is with some sort of strong bonded adhesive. And here is the observation, that we never see the broken vessel pieces leap together, reassemble themselves, and fly up into our hands completely whole again. That sort of thing doesn't happen naturally, if only it would. The Arrow of Time is set by the initial low entropy state of a system to a higher entropy state.[1] Okay, now that we have established the arrow of time we can see that time flows in a positive and forward motion with respect to the general consensus of the human

---

[1] Rigden, J. S. *MacMillan Encyclopedia of PHYSICS* (Simon & Schuster Macmillan, New York, 1996) Vol. 4 Pg. 1611

understanding. Some philosophers tend to argue that time is merely a way of expressing the relation between different events, a linguistic convenience. Just as the invention of a physical substance called *citizenship* was made to explain the relationship between Americans.

However, according to physics, there are some theories that suggest time travel to be allowed and some theories that forbid it. More specifically, time travel is theoretically allowed in the general relativity theory. But in superstring theory, time travel is forbidden compared to the general relativity theory. Stephen Hawking, coined to be one of the smartest men to have lived in modern day physics, mentions his concern for time travel because of the logical paradoxes. He posed the chronology protection conjecture, which secures protection of chronology by the simple statement that time travel is never allowed. One example of a logical paradox is this: Suppose time travel is allowed and you went back into time and made sure your parents never met changing the course of your reality, this would erase the outcome of you ever coming into existence. And if this could not be possible, and you are not allowed to even plan to go back in time and do such a thing, this would conflict with your free will to choose to do it in the first place. There are still a lot of unknowns about time and especially time travel. As a result, science has reached a dead end and hit its limits in this conversation. We conclude then that through physics, science has come to a great stand still about time travel. Nevertheless, I believe there are no physical limitations to higher dimensional precepts and higher dimensional beings such as spiritual creatures. Which makes me further contemplate about some spiritual beliefs Christians hold, myself included. With that being said, I'd like to share with you something that was birthed in my heart while thinking about time and time travel. I'd like to provoke your mind to think outside of the physical and limited world in which we live and for the next few minutes see the world through a spiritual lens. I don't claim for these ideas to be absolute truth but instead let these ideas inspire you to have greater faith in our all-powerful GOD and His great might! My prayer is for our faith to increase in the process.

Let's start off with this. What if some time in the future you did something wrong or you failed God and God being full of mercy and grace has brought us back to the 'today' so that we can relive it over again and have a better tomorrow? A better future? What if we are living in a 2nd chance that God has given us? What if we prayed and asked God sometime in the future for Him to give us a 2nd chance and to cancel or erase something we did? Could God not answer because He is limited to time? I believe that there is a deeper way to look at time. A way that we may not be able to fully understand.

So I asked myself if *prayer* was subject or bound by these limiting ideas and laws of time, and we know that God is not subject to time; He is not subject to the constraints of the fourth dimension. In fact, God is at a higher dimension. Can prayer therefore be subject to time? It is through prayer that I communicate to such a powerful God. So, can it be that prayer is subject and bound by these limiting ideas and laws of time? Or is my own understanding and position within the realm of time what really limits my prayer and communication with an almighty and limitless God. For instance, let's be participants of a thought experiment. As a Christian, I have learned the importance of prayer. My mother taught me to pray for everything. As is found in the Bible: the Lord Bless my going in and my going out.[1]

So let's say that you make it a habit to pray every day. You pray that God blesses you today! And after that you pray that God blesses you tomorrow. But you don't stop there and ask that God blesses you yesterday. You may ask yourself how can this be. How can I pray about my yesterday? That doesn't even make sense, within the constraints of our physical position and relation to time. Just follow me for a sec. Imagine that you make it a habit, a routine that every day you pray that God blesses you *today, tomorrow,* and *yesterday*. Watch what happens. Soon you will have a chain of blessings heading your way. Let's think about it this way. Let's say that for this example your prayers are made on a Tuesday and God hears your prayers and answers them immediately. (Of course if they are according to His will and are made in His Name,[2] the Name above every name, the Name of Jesus Christ) And those

---

[1] Deuteronomy 28:6
[2] John 14:14

answered prayers are packaged into a box and are shipped. This package has explicit orders from God to be sent directly to the address, your need. These can be prayers of protection, anointing, and favor. Prayers for God to give you wisdom, discernment, understanding, and gifts of the spirit. You could even pray that God may bless you indeed, that He may enlarge your territory, that God's hand may be upon you. Praying continuously that God may keep you from evil and that you may not cause pain.[1] And remember that every day you pray and send these prayers and blessings for; today, for tomorrow and for yesterday. That means there is a package of blessings and answered prayers that has been shipped to your *today*. Furthermore, there's a package full of blessings and answered prayers that you've shipped over to your *tomorrow*. And lastly, there's a special package full of blessings and answered prayers sent to your *yesterday*. Boom! The packages are all sent and are underway.

Okay, so far does that make sense? Remember that you are getting into a habit of doing this every day. So now, let's say Tuesday evening and night come and

---

[1] The prayer of Jabez found in 1 Chronicles 4:9-10

you go to bed like a regular day and wake up Wednesday morning. As soon as you wake up you brush your teeth and you dedicate some time to pray the same prayer. For God to bless you today, tomorrow and yesterday. As soon as you're done praying you get a package delivered to your front door. The packages is a box full of blessings and prayers from you prayers you shipped out that Wednesday morning. And you open it and are blessed. While you are still opening it the doorbell rings. You get up and open your front door and it's another package that has arrived and it's the package that you sent yesterday (Tuesday) when you did your regular routine prayer that morning. And now you have two boxes full of answered prayers and blessings. While you were still opening and being blessed by those two packages the doorbell rings yet again. You get up and walk to your front door full of anticipation and wonder. Guess what it is? It's another package full of answered prayers and blessings that your future self has sent you from Thursday. Your tomorrow-future-self did their regular routine prayer and shipped a package full of answered prayers and blessings your way that arrived today, Wednesday! Your future-self is praying for your current self. Just as your past-self was praying for your current self. Are you catching this? I truly believe in the power of prayer and habitual spiritual devotions. I've made this illustration here so you can visually see it and further wrap your head around this concept.

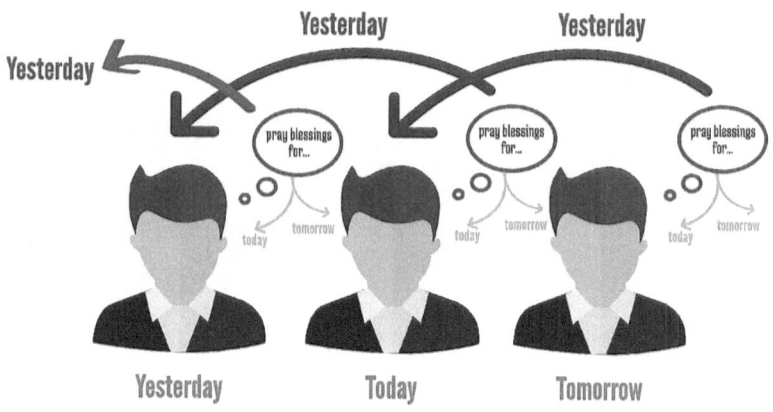

Can this be a way of traveling in time? Can this be a spiritual ability that we can tap into? I've heard of people saying they are covering their future children in prayer even when they're not married yet, nor do they presently have children. My older brother once asked me if I loved my children. I was 15 years with no job or even a girlfriend. Children were so far from my mind and from my present. Yet his question challenged me to start working hard now for my future children and family. See, I've heard about praying and thinking about the things that were to come but it was not until I met up with these truths that I began to understand the power of prayer. Prayer allows us to reach unimaginable places that may be impossible without prayer.

I've even heard of covering someone in prayer before but what if you cover them in different tenses: past, present and future. How many layers of covering could someone have? How many layers of blessings could someone obtain? This opens up an unlimited number of angles by which someone can be blessed through our prayers. Have we really been doing all that we can? Have we really been giving as much time as we could? When God calls us to pray, how deep does God want us to go? How much does God want to show us? What physical and natural limitations does God wants us to break in the spirit? There are no limits for God and He earnestly wants to take us there. God wants us to have breakthroughs in the spirit. Can you envision the possibilities we can have in the natural and in the physical if we first have these breakthroughs in prayer and in the spirit? Imagine the miracles that can flow through these breakthroughs? God is calling us for more! More of His presence and more of His supernatural encounters.

How does God interact with time? Does God experience a chronological flow of time? In this next chapter we will dive into God's timing; and what it means for the Almighty to be a part of time. Join me and put on your thinking cap once again as we delight in the meditations of our God!

chapter 3

"Could it be that the shine on Moses' face was simply the reflection of the face of Jesus on that day when Jesus took Peter, James, and John to the Mountain of Transfiguration?"

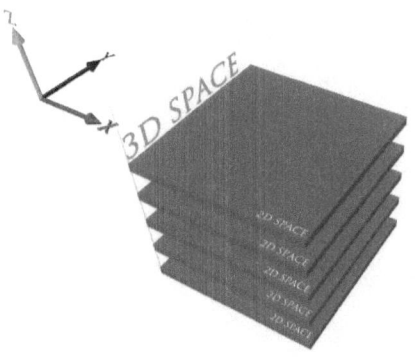

Judah Montenegro

God's Timing

Heli, a 30 year old man of God, finds himself in a horrible situation three years into his future where he desperately cries out to God needing a miracle. In the middle of his agony, pain, and suffering he goes deep into prayer where he gets ahold of the Lord's attention. He prays; "Lord, please let this cup pass before me, let these troubles disappear. Change my situation oh God. Shift things around so that I won't go through these sufferings." The Lord not only hears his prayers but chooses to answer them in a powerful way. Keeping in mind that Heli is three years into his future. In other words, Heli is praying this three years from now.

Then Heli wakes up one morning. It's no longer three years into the future but the calendar reads today's date. He gets up, does his daily routine and has a regular day. Some things start to happen in his life where he has no idea where it came from. Heli starts going through hard times here and there that make him wonder what's going on. His heart is steadfast in the Lord and his faith allows him to trust in God. He doesn't know that he is walking in a miracle. The things that he is going through are small compared to the things that the Lord has delivered him from in his future. Things like losing a job that will alter his upcoming decisions in life. Things like growing apart from a certain friendship. Things like an unexpected large bill or an unexpected trip out of town. God answered Heli's prayers by taking him back three years and making some life alterations in order to answer his prayers of the future.

I believe God can do this. Does God have restrictions when answering prayers? God can answer some powerful prayers in our future that start shifting things around now. Have you ever been living life when all of the sudden something weird happens? Something unexpected hits you. Something that should have not happened. I believe that these small shifts and alterations that happen in our life are answered prayers. Sometimes God knows that we need a miracle and He steps into our past from your future prayers to bless us today.

Imagine this was you, the future finds you going through a really tough situation where you cry out to God in prayer. In the middle of your pain and suffering you dive into prayer, deep prayer where you touch the hem of His garments.[1] Virtue leaves the master through your prayers moving the hand of God by changing your situation. So the hand of God grabs hold of you and brings you back a few years to the present, to the *now* so that you can have a second chance and not fall under those circumstances. God shifts things around and gives you a new hope because of His *everlasting* mercy! Thank God for His everlasting mercy and hope! His mercy endures forever! Now we can see all our circumstances as small and believe that everything is for our best. "What shall we then say to these things? If God be for us, who can be against us?"[2]

This is where you trust in God and have faith because God is working things out for your good because you love Him and are called according to His purpose.[3] And now you are set up to live a future with that answered prayer, you are living in a miracle! You may never know that you are living in a miracle. Imagine the miracles God has given us. Imagine the things that God has delivered us from. Car accidents, mistakes, falling into temptation, and

---

[1] Matthew 9:20
[2] Romans 8:31
[3] Romans 8:28

other "irreversible" occurrences. God is so good! Will we ever know ALL the prayers God has answered in our favor? What about all the unknown prayers of loved ones or our leaders, pastors, parents and mentors? What about when Christ prayed for you and I in the garden of Gethsemane right before He was to be crucified and killed for our sins and iniquities. Imagine that prayer. Imagine other prayers that were not recorded. Prayers that were made during His 40 day fast out in the wilderness. I believe Jesus had you and I in mind during these times. Where our Lord Jesus poured out His heart and laid down His life for you and me. He did this so we may be blessed and highly favored, that we might be protected from evil and that we may not fall into the plan of the enemy. I believe Christ was praying for you, for me. And He did this all out of love and all to have fellowship with you and me. That we may have access to the fullness of His presence His majesty could come and visit us. Oh Lord, who am I that you are mindful of me?[1] Who am I that you may know my name? Who am I that you would pray for my life and all my life's decisions? Take a moment to thank God for interceding for you when you most needed it.

---

[1] Psalms 8:4

In response to the question previously posed, Prayer has no limits! Prayer isn't bound to the laws of physics or the laws of nature! Prayer is released and freed into higher dimensions that are temporarily unknown to us. Prayer is free to transcend the space-time continuum. Time is not absolute and therefore is subject to that which is absolute. Time is relative to each individual. So if an individual steps into the depths of prayer, and/or higher dimensions of the spiritual world, time bends and is flexible to these higher dimensions that are not fixated nor restrained to time. Therefore, I believe Jesus had supernatural prayer encounters. By faith Jesus understood these truths and was able to tap into these supernatural encounters. Jesus practiced His faith daily and His faith failed Him not. Jesus spent time in prayer, sometimes all through the night.[1] He spent time learning the scriptures. Jesus spent time in the synagogues learning the scriptures of the prophets.[2] Jesus walked around with an unmovable faith. He was steadfast, immovable, abounding in the work of the kingdom of Heaven always, knowing that His work was not in vain. Jesus knew His work had a

---

[1] Luke 6:2
[2] Luke 4:16

purpose. He walked with such a confidence knowing that everything He did was the right thing to do and everything He said was the right thing to say at the right time. Because He understood the depths of the supernatural, the spirit and the higher dimensions.

In the gospels of Matthew, Mark, and Luke we see a time where Jesus takes Peter, James, and John up on a high mountain to pray when a supernatural encounter happens. The Transfiguration of Jesus on the mountain. Where His countenance began to shine like the sun and His clothes were exceedingly whiter than any earthly white. And behold there were two other men talking to Jesus, Moses and Elijah. They were talking to Jesus about His death that had not yet occurred in Jerusalem. And His disciples, Peter, James, and John all witnessed this supernatural experience. They were even frightened.[1]

Could this be the time where Moses, in the Old Testament, spends time with the Great I AM on Mount Sinai? Or perhaps the time when the Lord called Moses to the top of the mountain, and Moses went up.[2] Or when Moses drew near the thick

---

[1] Matthew 17:3 , Luke 9:28-30 , Mark 9:2-9
[2] Exodus 19:20

darkness where God was.[1] Or even when the Lord would speak to Moses face to face, as a man speaks to his friend.[2] What about the time when Moses spent forty days and forty nights and neither ate bread nor drank water.[3] Consider the time when he came down from Mount Sinai and was not aware that his countenance shone while he talked to the Lord.[4] Could it be that the shine on Moses' face was simply the reflection of the face of Jesus on that day when Jesus took Peter, James, and John to the Mountain? In the same way that the sun shines it's light so that the moon can reflect it. The Glory of the Lord then becomes such a place where one can perceive and enter into through prayer by faith. The Glory of God manifested through a supernatural experience. A glory accessible by invitation only. Fortunately for us, an invitation has already been extended by Jesus Christ to those who believe and accept Him as their Lord and Savior.[5] I believe the connection of these two moments that transcended the natural laws of the physical world was indeed a divine moment where

---

[1] Exodus 20:21
[2] Exodus 33:11
[3] Exodus 34:28
[4] Exodus 34:29
[5] Matthew 11:28 , Isaiah 45:22 , Hebrews 12:2

time became irrelevant. Two divine appointments connect, regardless of any physical regulations or natural laws that govern the heavens and the earth. When we further examine the time frame and occurrences of each, namely the Jehovah of the Old Testament from the experience as described in Exodus and the Jesus of the New Testament from the Transfiguration on the Mountain, we can conclude that Jesus is the same *I AM* as in the Old Testament.[1] And understand that Jesus is God; Jesus is the Lord of hosts. Therefore, Jesus and the Father are one.[2]

These divine appointments are what we call the Kairos time of God. Kairos time, by definition is the right, critical, or opportune moment in time. God's Kairos timing is perfect. Never late and/or never too early; it is precise. God's Kairos time is on a higher dimension that surpasses our limited understanding of time. As terrestrial beings our minds are limited to terrestrial thinking which includes chronological or linear timing and the time of this physical world as described by the early chapters of this book. It is hard to wrap our brains

---

[1] Exodus 3:14 , John 8:58
[2] John 10:30

around God's level of thinking when it comes to timing. God's timing is exact and without err.

What if Elijah wouldn't have been sensitive to God's Kairos time and never would have told Elisha to ask what he could do for him before he was taken up by the chariots of fire. Elisha would never gotten his double portion. What if Moses wouldn't have been obedient to the voice of God calling him to Mount Sinai? Moses wouldn't have gotten the Ten Commandments written by God's own finger. What if Peter, James, and John wouldn't have listened to Jesus when He invited them to the mountain to pray? They would never have been able to see that divine and wonderful moment where history met its maker and where Jesus talked with some of His friends.

Was Jesus showing Peter, James, and John that He had other friends with whom He could talk about the secrets that were to unfold. Was Jesus trying to inspire His apostles to keep diving into prayer so that they too wouldn't ever stop seeing Jesus? God is and will always long to have an intimate time with His children. He desires to have a time where He can share His secrets and have conversations. The kind of conversations that are good, deep and ongoing. Have you ever had a friend with whom you can share good, deep, ongoing conversations? Those kind of conversations that bless your soul and you can't have with just anyone. God yearns to have those types of conversations. God is looking for people to share His life with. God is searching for individuals He can spend time with and share His thoughts with. Individuals with whom He can have divine appointments.

Divine appointments come in all sorts, shapes, and sizes. They come when you least expect them. They may even come when you least want them. They can certainly come when you feel least prepared for them. God has a way of catching us off guard to see what we're really made of and to see our true intentions. Can you think of a time when you got caught in God's Kairos timing? Have you ever gotten lost on your way to an important meeting? You took a wrong turn or you got off an exit too early on the freeway. Those are God divine appointments. You must believe that the exit you took on the freeway was the exact exit God wanted you to get off at. Let's say God knows you have a special blessing of protection, kind of like an umbrella, that is covering you because you prayed that morning your regular routine prayer for God to protect you. And God wanted you to drive passed someone that needed that covering. What if that person you drove by was in the middle of suicidal thoughts and their soul was struggling to stay alive and God sent you there to that exact place in that exact moment. Without your knowing, God used you to aid that person's soul. Without the need to greet, pray over, or even speak a word to that individual their life was impacted and consequently saved. Simply by you driving by, the covering of protection in your life was able to preserve that God beloved soul. That is the Kairos timing of God.

I want to encourage you to be open and obedient to God's Kairos timing in your life. Don't let distractions come your way and separate you from God's perfect Kairos time. In the same way God has set divine appointments for you where he sets up your life events in such a manner that you make the exact decisions you need to make, so long as you are in the will of God, in order to complete God's perfect will and purpose. If we are completely surrendered to God's will, He will use us in ways we may never even know. No human could ever plan the way God plans.

God uses Chronos timing to develop and shape us into the people He needs us to be for His kingdom. God's Chronos timing is very important especially for us as human beings. This is because God wants us to be blessed and highly favored but He cannot bless and give us favor all at once. This would be too much for us to handle. Imagine you are a father (or mother) to a three year old child. You love that child so much and you want to bless them. So you take out your wallet and give them one thousand dollars cash. Then you ask this child what he or she wants to do with the money. They quickly respond that they want to go to their favorite candy store. You promptly arrive and let them buy whatever it is that they want. Their eyes pop out and they begin to buy all the candy they can lay their eyes on and get their hands on. I think you know where I'm going with this. The child soon will eat all this candy and it will be no good for their health and their future. Instead of the thousand dollars being a blessing, it became a curse. In the same way you would hand over the keys to a Lamborghini to teenager that just got a license. If the teenager has not been taught the proper weight of responsibility and all that it entails it will be too much for that specific time in their life, far more than they can handle.

In the same way, God loves us so much He wants to bless us and He loves us too much to give us things or blessings that are too much for us to handle. Remember God would never give us more than what we can handle.[1] I believe this is not limited to more blessing than we can handle. Because as we can see, a blessing given before we are prepared for it or before we can handle it can be a curse rather than a blessing. God means more than just good intentions in our lives.

---

[1] 1 Corinthians 10:13

PART II

"I used to think this was the beginning of your story. Memory is a strange thing, it doesn't work like I thought it did. We are so bound by time, by its order."

- Dr. Louise Banks, *Played by Amy Adams, in The Arrival, the movie*

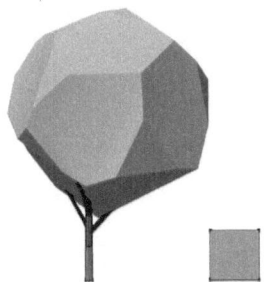

Judah Montenegro

chapter 4

"We can come together and be in one accord and open up the windows of heaven to set up a high spiritual dimension to be established in our midst."

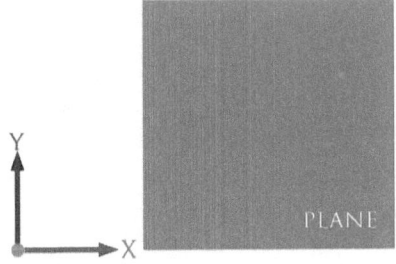

Judah Montenegro

Space & Matter

If you take a look around, you may see that there is a vast majority of space that is unoccupied by matter (matter; any sort of substance or material that occupies space and is perceptible to the senses in some way). Space is defined as a continuous area or expanse that is free, available, or unoccupied. If you recall in chapter one we briefly touched on time being described as a dimension. In this chapter, we want to zero in on and dissect the properties of space that make space three dimensions in a dimensional hierarchy. Whereby, the order of dimensions is set by nature.

A dimension is a measurable extension in a given direction, such as height, length, and width. Hence, three dimensions. A three dimensional (3D) world is composed of the measurable extensions height, length, and width. *Time, then,* becomes the fourth dimension. Time is a measurable extension in a given direction as we have established in Part I of this book. *Time,* being the protagonist thus far, will share the spotlight with the rest of the universe by becoming one with space. In which, we obtain Space-

Time; the space-time continuum. A subject that has puzzled physicists for decades. In Newtonian physics, referring to the work of Sir Isaac Newton, space, time, and matter are treated as quite separate entities. But in Einsteinian physics, space and time are combined into a four-dimensional continuum. And the two shall become one. Please help me welcome the beautiful couple Mr. and Mrs. Space Continuum! Einstein thought the two, space and time, were inseparable. These two are meshed and fashioned together because what God brings together, no man can separate. Can I hear an "Amen"?

These two are connected in a way where one can't exist without the other. Time moves forward with space. Think about it this way, a plant lives in space but is not independent of time. The plant grows and flourishes WITH time. Do you see that? Any object in space passes with time. An ice cream cone, taking up space, melts WITH time. An apple having a height, length, and width decomposes WITH time. Do you see how these two, space and time, are fastened together? These two combined become the star of the universe, pun intended. Making them a dynamic dual, the perfect pair, a marvelous masterpiece.

But what if I told you there are more dimensions than that? What if I told you that mathematicians have discovered more than 11 and 12 dimensions? Of course these dimensions are possible in mathematics by brut mathematical equations and heavy computational arithmetic. But what does that mean to a physical world? How can we understand this and what use do we have for it. Allow me to break down the dimensional hierarchy in order to understand higher dimensions and the invisible world that the Bible calls more real than the visible world.[1] The purpose to understanding these truths is to grasp the depths of the unknown world that the bible calls the spiritual world or the Kingdom of Heaven. I believe as a Christian and follower of Jesus that there are a lot of truths unknown and unseen to me but just because I can't see it or just because I don't understand it doesn't mean that it doesn't exist.

Let's put on our thinking caps and prepare our imaginations to visualize these truths as they exist. We'll start off real basic and simple in order to build up and understand the more complicated concepts of higher dimensions. Although, some dimensions are

---
[1] Hebrews 11:3

very simple to understand and visualize while others are a little more abstract and not so intuitive. Nevertheless, here we go, stay with me!

We'll start off with the zero dimension. A point. A dot. It has no measurable extension of any sort and has no range of direction. It has zero dimensions.

● DOT

Then we have the 1st dimension. A line. A line, has one measurable extension. It has one range of direction. You can measure a point on a line. Whether it be two or three ticks away from the center of the line in either direction. It can be represented by length, for example.

LINE

Next we have the 2nd dimension. A plane; A flat or level surface. An area or surface having determinate extension and spatial direction or position.[1] Here we have our regular sheet of paper. Our desk top or kitchen table surface. Our wall surface, etc. Here we have two measurable extensions and two ranges of direction. If you've ever played BattleShip you know how the coordinate system works. In elementary school we learned how to read coordinates. Essentially, we were learning positions of a point or positions of a curve on a two-dimensional graph.

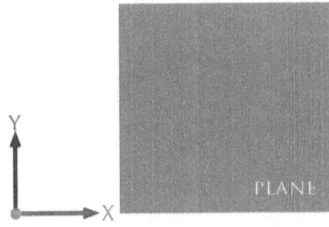

The following is the third dimension. I say this is the most famous dimension of them all. It gets everyone excited when they hear 3D movie experience or a 3D printer and so on. 3D is well know because it is easy to understand, see, and touch. When you think of a Rubik's Cube or a cube of sugar or even a shoe box. You are seeing a three dimensional object. It has height, length, and width. These three

---

[1] Dictionay.com

dimensions are visible and relatable because we deal with them on an everyday basis. Our brains were made to understand three-dimensional space. Even our memory was made to store information in a 3D fashion. Our whole life and way of thinking is in 3D.

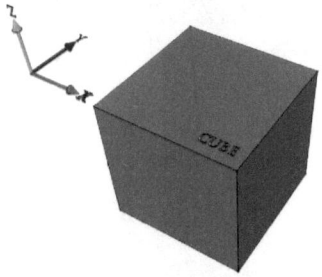

Then we have the fourth dimension which was discussed a bit at the beginning of this chapter. The four dimensions here are height, length, width, and time. These four are measurable extensions with given directions. The first three dimensions are spatial dimensions and the last is unidirectional, as we have established time only flows in a positive direction.

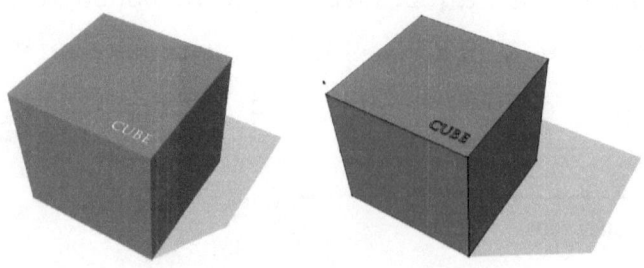

As seen by the two cubes that the shadow grows or changes in shape due to the light source moving or changing with time. Just as the sun would change its position as time passes.

Then we can go into the 5th and 6th dimensions and beyond our physical understanding to where it ceases to comprehend the essence of these higher dimensions. I believe that these higher dimension can be different levels of the spiritual realm. The spiritual realm can be divided into different levels as we will be able to see as we expose the structure of these dimensions. I'd like to go on a dimensional journey starting from the zeroth dimension and so on. Join me in exploring deeper into these concepts.

Now that we know what the basic dimensions look like we can see how they are organized and stacked in order to form real world dimensions.

Starting with the zeroth dimension, a point. Suppose this point is the only thing in the whole universe the only thing around. It is its own individual identity, essentially, its own universe, its own world. Along comes another point having its own individual outlook on life and respective journey. These two points meet and have so much in common that they become best friends and never leave each other's side. They are juxtaposed. They are now aligned and are in one accord, living in harmony. What has happened here? If you notice carefully here you can note that they have set themselves up to be a one-dimensional entity. Let me explain.

Take a line, in the 1st dimension, is made up of infinitesimal number of points all juxtaposed. All these points, when aligned, form the next dimension above. If you were to draw on a sheet of paper one hundred points aligned, right next to each other, you would begin to see a line. Which by definition is the 1st dimension. Now, follow me. If you would take the same story and apply it to a one-dimensional element what would happen?

A line is living all alone in a universe when all the sudden you introduce another one-dimensional element, a line, and juxtapose it with the other line that is already there. They become best friends and are in one accord, they live in harmony. What happens? They have set themselves up to be a two-dimensional entity. Take a pen out and draw one hundred lines that are juxtaposed. What do you notice? You have made a two-dimensional element. A plane. Same thing happens when you have multiple planes in juxtaposition. It creates a third dimensional entity. Let me illustrate it this way. Think of a sticky note. It's shaped as a square. That sticky note is a two-dimensional plane. When you put another one on top of it you create a thicker plane. Then you place a few more on top and it gets thicker. Say you have one thousand sticky notes on top of each other, congratulations, you have created a third dimensional element.

We see how lower dimensions can come together and be aligned in such a way that sets up the next higher dimensional entity. Could this be possible when we think about us? We can come together and be in one accord and open up the windows of heaven to set up a high spiritual dimension to be established in our midst. Could this be what the scriptures were talking about when they say "For where two or three are gathered together in My name, I am there in the midst of them." in Matthew 18:20?

In this next chapter I want to prove this. And I'll show you the logic and the physics behind it with scriptures to support my findings. Hurry over and don't try to take off your thinking cap. While you're at it, grab your spiritual cap or your spiritual eyes and follow me on this journey.

chapter 5

"How much then, do we really know about God if we are only exposed to the shadows of his greatness?"

 LINE

Judah Montenegro

Lord of the Dimensions

Before we jump into understanding the different dimensions, I want to be clear and mention that I don't believe God is simply a higher dimensional being. I believe that our GOD almighty is the Lord of the Dimensions and He can enter, exit, exist and remove Himself from any and all dimensions simultaneously whenever He pleases. He is the Lord of the Universe and can do and be as He wills. Our God has no limits.

Okay, good! Now that I cleared that up, allow me to show you how a four-dimensional entity sets itself up for a higher dimensional exposure. Imagine a 3D object flowing through the space-time continuum. It makes its way in the arrow of time leaving a linear record of its passing in the catalogs of time measurements. In other words, we notice that the arrow of time only flows in one direction. I consider that to be a linear direction. So, when looked at from a higher dimensional point of view, the 4D element is a linear entity. Why linear? This is because space-time is unidirectional, flowing in the forward direction of time. Time cannot flow backwards nor from side to side. It can only flow in one measurable extension or one specific direction. That makes it a linear construct. What happened the last time we had a linear construct also known as a line? We juxtaposed the bad boys and set ourselves up to establish the next dimensions in sequence. Therefore, by juxtaposing two or three four-dimensional elements we create a fifth dimensional entity! Boom! We have just established a 5th dimensional object simply by using logic.

We may ask ourselves how is it that you juxtapose a 4D element? Do you just place it side by side like you would do with 3D objects? Well the truth is that it's not that simple. I believe that you juxtapose a couple of 4D entities by aligning both of their purposes. By bringing the two, or more, 4D elements together in one single purpose or vision. When two or more 4D elements are aligned and in purpose and in vision the two become one in a higher dimension. Hence, setting itself up for a higher dimensional exposure.

So now that we have a fifth dimensional element we can juxtapose a few of those through purpose and vision and compose a 6th dimensional construct. We can repeat the process and create a loop! A loop of never ending higher dimensions! How great is our GOD! Wow, I am blown away by how cool this stuff is. I am blessed to know that God is the Lord of all dimensions. And I get a chance to call Him my friend![1]

I believe that as humans and creations of God, we are not just four-dimensional beings. I believe that we are 5th or even 6th dimensional beings because we have a soul and a spirit. Our spirit is awaken when we give our lives to the Lord and undergo a rebirth in our lives through the baptism by water[2] and by fire.[3] This rebirth takes place when we firstly, repent from our sins and are baptized by water as a public testament for all to see, and secondly, by fire through the gift of the Holy Spirit by speaking in new tongues.[4] What if baptism by water is the baptism the 3D part of our life needs? I'm referring to the flesh, the body and the bones; the physical elements of our life. Then baptism by fire, through the Holy Spirit and speaking in tongues is the baptism required for the

---

[1] John 15:15
[2] Acts 2:38
[3] Matthew 3:11
[4] Acts 2:4

4D part of our life. That is, the mind, the soul, our consciousness and subconsciousness. Isn't that something?

I believe that we have access to the almighty God through the spiritual realm, as a result, this makes us 5D beings. The spiritual realm is the higher dimensions. Of course God can enter any level or dimension at any time, however and whenever He so chooses.

Allow me to give you an illustration of higher dimensions. I will use lower dimensions as similes and analogies in order to get my point across. Imagine a "flat-land being". An individual that lives in a two dimensional plane. This "flat-land being" can walk around this 2D plane as it pleases. It goes inside its little square home, shuts the door and cannot see outside its home. This "flat-land being" walks up to different objects and elements in its 2D world but can only see the outer perimeter of the object or element. It cannot see the inside.

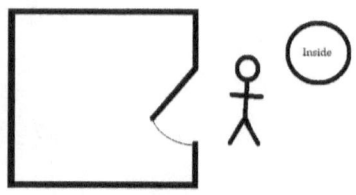

If you get a 3D object and lay it on top of the 2D plane, the 2D individual will only see the outer perimeter of this 3D object that is making contact with the 2D plane. Say the 2D being is living on a sheet of paper and you get a pencil (a 3D object) and poke it right through perpendicularly.

The individual living on this 2D sheet of paper will only see and notice the thin slice, or cross section, of the pencil that is making contact with the 2D plane. As you poke the sheet of paper with the sharp point and move towards the middle of the pencil, the circle, the cross sectional perimeter that the 2D being is seeing, gets larger.

Carl Sagan made a very cool video that breaks down higher dimensions where he illustrates it very clearly. I encourage you to check it out. It's on YouTube under *4th Dimension - Tesseract, 4th Dimension Made Easy - Carl Sagan.* I really enjoyed watching this video. Matter of fact, when I watch videos like this or even learn more about higher dimensions my faith is built and I grow closer to the Lord. I get to see biblical truths be spelled out with physics and mathematical terminology. I feel that my spiritual understanding grows and I am able to be a sound Christian. I mean think about it, we are seriously breaking down these complex concepts and applying them spiritually to understand more about our God and the kingdom of heaven. Let's continue to use the lower dimensions and that same language as analogies to think about our higher dimensions.

Since we are 4 or 5 dimensional beings we only see the "perimeters" or "cross sectional views" of a higher dimensional manifestations of God. The same way the 2D being can't see *inside* a circle because they don't have the capability, we do not have access to greater understanding beyond the perimeters in which we are permitted to operate. This is not because they/we are ignorant or not smart enough, rather they/we just are not able to because it exceeds the capability of our dimension. They/we are limited by their/our circumstances and actual state of being. The higher dimensional being looking into a lower dimensional world can see inside lower dimensional closed structures. For example, us as higher dimensional beings, compared to the 2D being, are able to see inside and outside the square 2D house simultaneously. In the same way, this is how GOD, being a much higher dimensional entity, can see our heart.[1] He can see the depths of our soul and spirit. He knows our thoughts before we think them. He knows the words before we even pronounce them. GOD knows all our ways. We cannot hide from Him. Our literal darkness cannot make us disappear nor does it make us become unseeable from His sight. For the night time is as the bright and clear as the noon day to Him. Such knowledge may be too much for us to fathom.[2]

---

[1] 1 Samuel 16:7
[2] Psalm 139

Get this, when we see miracles, signs, and wonders we are only seeing the shadow of a much higher dimensional being at work. Our GOD almighty is at work and since we are not at His level, since our thoughts are not His thoughts, and since our ways are not His ways[1] we can only perceive the mere shadow of what is happening. Just as a 3D pencil casts a shadow on a 2D world, God, being a higher dimensional being casts His shadow on us lower dimensional creatures. Or that same way in which a 3D pencil, with aid, can draw and add or subtract figures that become substance on a 2D world, GOD can come in and add or subtract figures and substance to our world. We may speculate and try to understand what is occurring, however, we cannot thoroughly understand the ins and outs of the what is actually taking place when GOD is at work. Therefore we call it a miracle or a wondrous manifestation.

---

[1] Isaiah 55:8-9

Do you see the relation? Higher dimensional beings and entities can be manifested onto lower dimensional worlds. Hence, we have heard sightings of angels and angelic demonstrations. Strange occurrences where the angle is there one moment and then it's gone. Maybe you have had an angelic experience yourself. Or perhaps even the opposite, the appearance of evil spirits and demonic possessions. We read of such events in the Bible, when Jesus encounters the man with many unclean spirits inside of him.[1] But Jesus has compassion over him and heals the man from the unclean legion of spirits. What unseen doors did this man open for these unclean spirits to take part in his life? Lord, help us not leave any higher dimensional doors open for unclean spirits to come visit and dwell in us. I urge you to ask for forgiveness if at any time you have, willfully or inadvertabtly, consciously or subconsciously, opened any doors for the enemy of our souls to come into your life. If we repent and let God be the Lord of our lives He will readily inhabit our being and will leave no room for any other

---

[1] Mark 5:1-20

dwellers. He is a jealous God and He wants our whole heart.[1]

When we see things that are irregular in our world we are only seeing a glimpse or just the shadows of what a higher dimension is casting. Imagine the secrets and depths of a world we cannot see? Personally, I want to be on the good side of those depths. I want to abide under the shadow of the Almighty.[2] I find it very interesting that now we can see scriptures and look at them from a different point of view. For instance, when Peter would walk and his SHADOW would heal the people.[3] Was it really his shadow? I think not. It was the power, the virtue, of a higher dimension that was working "behind the scenes." We call that higher dimensional power in Peter the Holy Spirit. The observers were only able to see the shadow of what was really happening and recorded it in such the way we read it now.

How much then, do we really know about God if we are only exposed to the shadows of his greatness? This is mind blasting because God has so

---

[1] Deuteronomy 6:5, Matthew 22:37-38
[2] Psalm 91:1
[3] Acts 5:15

much more depth to Him than we can ever fathom! We have not even scratched the tip of the iceberg, we have only been exposed to the shadow of the almighty!

Let's keep looking at some more instances where we get a glimpse of higher dimensions in the bible. There's a portion in the bible where the apostle Paul is writing to the church in Corinth:

> "It is [a]doubtless not profitable for me to boast. I will come to visions and revelations of the Lord: I know a man in Christ who fourteen years ago – whether in the body I do not know, or whether out of the body I do not know, God knows – such a one was caught up to the third heaven. And I know such a man – whether in the body or out of the body I do not know, God knows – how he was caught up into Paradise and heard inexpressible words, which it is not lawful for a man to utter. Of such a one I will boast; yet of myself I will not boast, except in my infirmities."
>
> *2 Corinthians 12:1-5*

Here the words "the third heaven" has a connotation in the greek meaning the heights above, the upper regions. And is found in Strong's Exhaustive Concordance here; Perhaps from the same as *oros* (through the idea of elevation); the sky; by extension, heaven (as the abode of God); by implication, happiness, power, eternity; specially, the Gospel (Christianity) -- air, heaven(-ly), sky.[1] And the word "paradise" has a connotation from the greek Strong's Exhaustive Concordance; Of Oriental origin (compare pardec); a park, i.e. (specially), an Eden (place of future happiness, "paradise") -- paradise.[2]

This leads me to believe that the Apostle Paul could have been talking about himself. Was he taken up to a future or a past where he was able to enjoy the blessings of the Garden of Eden? Could he have been taken to the future dwelling place of the saints? Was he taken into a whole new other dimension? Did God allow them to transcend into a higher dimension? This I do not know, only God knows. But I believe it could have been a combination of both. Paul could have been taking into some other time frame through

---

[1] Biblehub.com/greek
[2] Biblehub.com/strongs.htm

a higher dimension and the only way he could have described it was an out of body experience, or an out of the regular 3D and 4D experience where he heard "inexpressible words."

There's another occurrence in the bible where Jesus heals a blind man in Bethsaida. It's found in the book of Mark chapter 8. Check it out.

> *"Then He came to Bethsaida; and they brought a blind man to Him, and begged Him to touch him. So He took the blind man by the hand and led him out of the town. And when He had spit on his eyes and put His hands on him, He asked him if he saw anything. And he looked up and said, 'I see men like trees, walking.' Then He put His hands on his eyes again and made him look up. And he was restored and saw everyone clearly."* Mark 8:22-25

Now, I don't think Jesus made a mistake and didn't heal him properly the first time. The son of God wouldn't be prone to such silly errs. I believe that Jesus allowed him vision into a higher dimension. Maybe the same dimension where Jesus sees the heart of men or the fruits of men. In the book of Psalms chapter one we see men being compared to trees. "...he shall be like a tree planted by rivers of water..." In John chapter fifteen it refers to us, followers of Jesus, as branches that must *bear fruit*. In John chapter seven it talks about knowing men by their *fruits*. We see these comparisons to trees with fruit. I don't know if the man was blind from birth but what if he was? He'd never seen a tree before. How did He know what a tree looked like? Maybe just by feel but I believe Jesus gave him a glimpse into a deeper dimension where he was able to see men the way God sees them. Within a dimension where it is clearly visible whether man produces good fruit or bad fruit. Oh Lord help us! I believe God can see our fruits very clearly. In the same way God can see the intentions of the heart and the thoughts like we can see leaves on a tree. God give us grace and favor.

Another portion in the bible where Elijah was basically asked not to disappear and be taken by the spirit of the Lord into an unknown place.

> *"As the Lord your God lives, there is no nation or kingdom where my master has not sent someone to hunt for you; and when they said, 'He is not here,' he took an oath*

from the kingdom or nation that they could not find you. And now you say, 'Go, tell your master, "Elijah is here" '! And it shall come to pass, as soon as I am gone from you, that the Spirit of the Lord will carry you to a place I do not know; ..." 1 Kings 18:10-12

Perhaps Elijah had a reputation of being taken by the Spirit of the Lord into unknown places. Maybe he would disappearing and go to a place where people couldn't see him. The Spirit of the Lord may have taken him up into a higher dimension to spend time with the Almighty or travel in time to visit the New Testament. The bible doesn't record all the times that Jesus had powerful prayer encounters with His buddies. Maybe Elijah was a regular and would disappear and go hang out with Jesus in the New Testament. You know, God may have access to wormholes that allow you to travel through the space-time continuum. God may have access to some sort of portal that allows you to travel into higher dimensions, I mean, why not? He is the everlasting GOD, the great I AM, the author and finisher, the Alpha and Omega, the beginning and the end, the one who was, the one who is, and the one who is to come![1]

---

[1] Revelations 1:8

Judah Montenegro

chapter 6

"I can connect to my older, wiser, more experience future-self and receive mentorship from the Holy Spirit in my future-self."

Judah Montenegro

Inspired

An individual walks into your workplace and you see them at a distance. Their shoes are polished and their clothes are pressed. They look awfully familiar and you can't put your finger on how you know them. Their hair is tidy and they walk with a certain confidence. They have great posture and a clean cut finish to their step. You notice such class from this individual that is very rare to see. They begin to talk to some of your coworkers but you can't hear what either of them are saying. You surmise the conversation must be exceptional because it appears your coworkers are enjoying the interaction while showing such high regard for this individual. You gather their vocabulary is impeccable and their wit is delightful to the ears. Finally, this individual takes a seat and begins to work with such diligence and grace. Who is this? They seem so familiar; where have I seen them? You ask yourself. When all of the sudden you are floored by the fact that it was YOU the whole time. You stand up and begin to put the pieces together. It was a vision. You were seeing yourself. A better version of yourself, refined and elegant.

What happened here? The Lord gave you a glance at what your future holds. Have you ever wondered what your future self is like? Well I want to encourage you to explore this notion that you can literally design that individual detail by detail. It is as if you were to walk up to an individual and pick and choose the characteristics you want from them. You can take away all the bad habits and add all the qualities you want and the attributes your loved ones need from you. Try it. Take out a paper, draw your future self. Draw the kind of smile you want. Picture the kind of hairstyle you want. Sketch an awesome outfit with the best waste line. Paint a picture of your favorite pair of shoes. As a matter of fact, you don't have to be an artist to design this *you*. You can illustrate with your words by making a list or writing out a paragraph. You can create an individual that would be awesome to wake up as tomorrow morning. Write down what type of food they eat and what type of friends they have. Specify the connections to which they have access and with whom they rub shoulders. Document their type of car and the house in which they live. Note their exact weight and health habits. Use as much detail as your imagination can produce. Then submit it to prayer! Look at this drawing every day. Read that list or detailed description daily until you can practically see that person walk into your workplace. Ask God to bless you with that future! Ask God to help you become that individual. Claim it, declare it, believe that it's yours and it shall be! The Bible tells us where there is no vision, the people perish.[1] Have a vision and know exactly what you want and need.

I want to encourage you to have a vision. You can do the same for your career, dream house, spiritual goals, and whatever other aspirations you choose to pursue. I encourage you to always seek first the kingdom of God and His righteousness and all else will be added.[2] Be steadfast immovable, abounding in the work of the Lord always, knowing that your work in the Lord is not in vain.[3] Hide these words in your heart that you man not sin against the Lord.[4] Then, when times get tough you have to press towards the mark for the prize which is in the high calling of our Lord Jesus Christ.[5]

---

[1] Proverbs 29:18
[2] Matthew 6:33
[3] 1 Corinthians 15:58
[4] Psalm 119:11
[5] Philippians 3:14

There will be times where you will need encouragement. You will need mentorship. You will need someone to look up to. Someone to speak words of life into your heart and soul. There will be times when you feel like there is no one you can look up to or no one to mentor you. And that is when you learn to encourage yourself. You learn to preach to yourself. You learn to believe that you can speak life into your situation and cast out negativity. You cast out evil spirits that come your way, cast out spirits of depression and spirits of low self-esteem. You learn to rebuke spirits of lust, evil temptation and addictions, in Jesus Name!

When all else fails call upon the Holy Spirit to come and guide you and mentor you and be your best friend in times of loneliness in the wilderness. I have learned that because I am a born again, believer, child of God I have the Holy Spirit in me. The Holy Spirit in me today allows my past self and my future self to have it also. Therefore, if my future self and my present-self have the same Holy Spirit, I can connect to my older, wiser, more experience future-self and receive mentorship from the Holy Spirit in my future-self. Did you catch that? That means I have access to a better version of myself *now*. Get this, I can mentor my present-self from my future-self through the Holy Spirit. And in reality it's not me doing the mentoring and guiding, it's the Holy Spirit. What a blessing!

Be inspired by your future-self *now*! Step into the dimension of the Holy Spirit through prayer to travel in time and mentor yourself with your very own guiding Holy Spirit! There are sometimes when I needed some mentorship and guidance and couldn't find it anywhere else but in my prayer time with the Lord. Believe with me that He is our provider. If you're following me, I am saying we have access to time traveling in prayer, we have access to higher dimensions through prayer. And we can achieve all this when we spend time with GOD.

Now I do want to say this; I don't believe that we are independent beings that can live off of self and future self. I am not saying that we have the answer within ourselves and all we need to do is to search deep within ourselves. No, I am saying that the same power that conquered the grave and rose Jesus up out of that tomb after the third day lives in us.[1]

---

[1] Romans 8:11

That's the Holy Spirit in us. If we tap into that power and into those depths of the spirit we can conquer anything! Because I can do all things through Christ who gives me strength![1]

When the Apostle Paul writes that he knew a man in Christ fourteen years ago that was taken up into the third heaven? I'd venture to say that he met his future self in Christ, through the Holy Spirit! He was caught up in a higher dimension which allowed him to encourage himself or sow a seed of life into his subconscious, into his spirit. The Apostle Paul traveled in time or was lifted up into a higher dimension where he broke the laws of physics and nature to receive a powerful gift from God. He heard unspeakable words that cannot be repeated. Imagine what that means. He saw things that can't be explained with human words or human understanding. Since we process all information in 3D format we are not able to comprehend higher dimensional information. Our eyes and ears and senses cannot record such things. Our memories only record in three dimensions. Our understanding is all computed in a 3D or lower fashion. That's why the

---

[1] Philippians 4:13

GOD almighty undressed Himself of Royal garments and put on earthly limits. He submitted Himself to a simpler dimensional life in order to show us his unconditional LOVE. GOD humbled Himself to the point of death, and death of a cross. And He did this all because He loved us so much He wanted to save us from eternal damnation. God wants us to have life and life in abundance, a life of heavenly glory. God wants to take us to His heavenly dwellings[1] which are found in those upper dimensions of His!

I want to leave you with a letter my future-self wrote to my past-self, which was at that time, my present-self. That way you are inspired to think outside of the third and fourth dimension into the blessings God has for you.

Dear Judah,

The truths that I will leave you here in this book are for you to read in the depths of your spirit and to analyze and to get them on paper and soon after publish this book!

---

[1] 2 Corinthians 5:1

Please, don't wait any longer. The time is now! The Longer you wait the longer you separate yourself from your blessing and your true purpose in the blessed life that the Lord has given you!

Write it Judah, even if you don't get it all. It will make sense as you write and as you become an author.

Don't let this world or the desires of your physical-3D world stop you from reaching your higher dimensional purpose!

I Judah Montenegro, write to you Judah Montenegro in the Future that the Holy Spirit in you now will be awaked to me here in my time. The intention is that I may be blessed, inspired and feel the discipline that I need to reach my purpose! Please, Holy Spirit of my future, bless me now that I may step into my calling and my purpose and not waste time! I don't want to live a life full of regrets or wasted time!

Holy spirit of my future, help me learn the organizational skills and disciplines I need now to move forward with this gift of truths you have given me. Lord help me!

I search for inspiration here on earth in my now and I find none! There is no man that I look up to whole heartedly. There is no man that has all the qualities that I want to have to be a solid man of God. A man who has mastered the balance of life in all areas. All areas such as: Health, Growth, maintenance in; spiritual, mental, emotional, and physical health and even the areas that I have yet to know.

Since I can't find a mentor or an inspirational being that I can look up to for all these areas of my life I look to my future Holy Spirit to inspire me and teach me and give me advice. My prayer is: Lord, release early, release now, what You have already given to me. I request that I may accomplish early, now, what the You have already delivered into my hands. I submit to You that I may tap in to that unlimited source called the Holy Spirit and be able to obtain wisdom, knowledge, and understanding to be a blessing to my surroundings early, now! In Jesus Name.

I believe God is letting me reach these truths so that the enemy may be reminded, that he is already defeated, now!

I write to you from the future! Letting you know that this book is going to be a hit! The knowledge you are releasing is going to be appreciated by so many hungry individuals that are ready to read and learn what you have to share. You have been holding these truths inside of you for quite some time but the time is now to share and open up the windows of what you have. Don't be afraid. Don't be scared of what people might think. People are not thinking about you in the past so much as to affect your future; in fact, they have nothing but gratitude towards you because you had the courage to publicize this book in the future. But this can only be so if you release it now! GO! Write! Think! And be free!

Signed,

Judah Montenegro

February 2016

I encourage you to take this example, something that pushed me and urged me to live out my future and better self, and make it your own. Write to yourself, declare the blessings you want to see for yourself now and watch them as they become your reality developing into your new purposefully designed existence.

The End

or the beginning...

THE SCIENCE OF
FAITH

www.ingramcontent.com/pod-product-compliance
Lightning Source LLC
Chambersburg PA
CBHW031434210526
45464CB00005B/2196